USS MIDWAY AIR WINGS

A pictorial chronology of Midway's air groups and air wings

By CDR Pete Clayton, United States Navy

74-B22

I0642992

VB-74 SB2C-5 *Helldiver* flying from *Midway*, c. March1946.   *National Archives*

TABLE OF CONTENTS

VBF-74 *Corsair* parked alongside *Midway*'s island shortly after commissioning. The original island configuration was unworkably cramped and lasted less than a year. *National Archives*

Vought F4U-4 *Corsair* taxis forward out of the pack aft in preparation for deck launch during shakedown cruise to the Caribbean, 21 November 1945. This aircraft is assigned to VBF-74. Together VBF-74 and sister squadron VF-74 brought 54 *Corsairs* aboard *Midway* for her shakedown cruise. *National Archives*

CVBG-74
7 November 1945 – 12 December 1945
Shakedown - Caribbean
CO - Captain Joseph F. Bolger
CAG - CDR John T. Blackburn

VF-74	F4U-4/F6F-5P	74-F-xx
VBF-74	F4U-4	74-BF-xx
VB-74	SB2C-4E/SB2C-5	74-B-xx
VT-74	SBW-4E/SB2C-4E	74-T-xx

Commander Air Group-74
CDR John T. Blackburn

A VA-1B SB2C-5 commences deck launch, c. 1946. The SB2C-5 was the final version of the Curtiss *Helldiver*. It was in effect an SB2C-4 with increased fuel capacity. More than 7,000 SB2Cs were built of which 988 were SB2C-5s. *Helldivers* operated from *Midway* from commissioning until 1947. The *Helldiver* was powered by a Wright R-2600-20 Cyclone radial engine producing 1,900 hp. which yielded a maximum speed of 294 mph. It was armed with two 20mm wing-mounted cannons and two 30 cal. machine guns in the rear cockpit. It could carry 2,000 lb of bombs or one Mk 13-2 torpedo in the internal bay. Up to 500 lbs of bombs could be carried on each underwing hardpoint.

CVBG-1
29 October 1947 – 11 March 1948
Mediterranean
CO - Captain Albert K. Morehouse
CAG - CDR R. Emmett Riera

VF-1B	F4U-4B/F6F-5P	M 1xx
VF-2B	F4U-4B	M 2xx
VA-1B	AD-1	M 3xx
VA-2B	AD-1/TBM-3W	M 4xx
VCN-2	F6F-5N	LA xx

Commander Air Group One
CDR R. Emmett Riera

Douglas AD-1 *Skyraider*, one of 27 assigned to VA-2B, commences a deck launch run in the Mediterranean, 13 January 1948. *Skyraiders* operated from *Midway's* deck from 1947 through the Vietnam combat cruise of 1965, a period of 18 years. The AD was powered by a Wright R-3350-24W Duplex Cyclone 18-cylinder radial engine, rated at 2,500 hp for takeoff. Its maximum speed was 366 mph at 13,500 feet with an initial climb rate of 3,590 feet/min. The *Skyraider's* service ceiling was 33,000 feet and it had a range was 1940 miles with 2,000 lbs. of bombs. It weighed 10,500 lbs. empty or 18,000 lbs. fully loaded. It was armed with two 20mm cannons in the wings. The first AD-1 flew 5 November 1946. 277 were built. *National Archives*

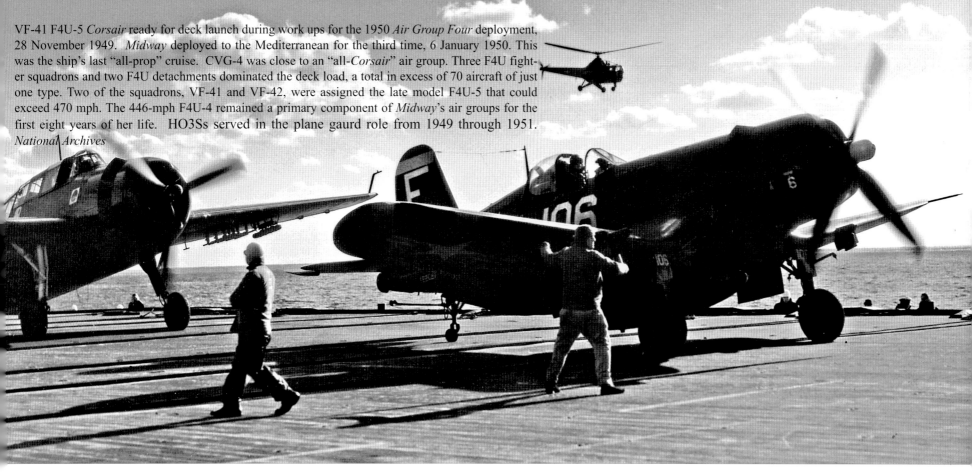

VF-41 F4U-5 *Corsair* ready for deck launch during work ups for the 1950 *Air Group Four* deployment, 28 November 1949. *Midway* deployed to the Mediterranean for the third time, 6 January 1950. This was the ship's last "all-prop" cruise. CVG-4 was close to an "all-*Corsair*" air group. Three F4U fighter squadrons and two F4U detachments dominated the deck load, a total in excess of 70 aircraft of just one type. Two of the squadrons, VF-41 and VF-42, were assigned the late model F4U-5 that could exceed 470 mph. The 446-mph F4U-4 remained a primary component of *Midway*'s air groups for the first eight years of her life. HO3Ss served in the plane gaurd role from 1949 through 1951.
National Archives

CVG-6		
4 January 1949 – 5 March 1949		
Mediterranean		
CO - Captain Marcel E. A. Gouin		
CAG - CDR Grafton B. Campbell		
VMF-225	F4U-4	WI xx
VMF-461	F4U-4	LP xx
VF-172	F4U-4	R 2xx
VA-64	AD-1	C 4xx
VA-65	AD-1	C 5xx
VC-4 Det 5	F4U-5N	NA xx
VC-12 Det 5	TBM-3W	NE xx
HU-2 Det 5	HO3S-1	UR xx

CVG-4		
6 January 1950 – 23 May 1950		
Mediterranean		
CO - Captain Wallace M. Beakley		
CAG - CDR Richard H. Burns		
VF-41	F4U-5	F 1xx
VF-42	F4U-5	F 2xx
VF-43	F4U-4	F 3xx
VA-44	AD-1	F 4xx
VA-45	AD-1	F 5xx
VC-4 Det 5	F4U-5N	NA xx
VC-12 Det 5	TBM-3E	NE xx
VC-62 Det 5	F4U-5P	PL xx
HU-2 Det 5	HO3S-1	UR xx

Commander Air Group Four
CDR Richard H. Burns

Vought XF7U-1 *Cutlass* aboard *Midway* for Naval Air Test Center carrier suitability trials in July 1951. The *Cutlass* first flew 29 September 1948. The F7U-1 was largely betrayed by its anemic J34 Westinghouse turbojets, engines which some pilots wryly observed put out less heat than the same company's toasters. Engine response for latter stage wave offs were deemed unsatisfactory. Its carrier handling was notoriously poor.

The F7U-3 with more powerful Westinghouse J46-WE-8B turbojets eventually became the fleet version. 288 aircraft were assigned to 13 Navy and Marine squadrons. Further development stopped once the F8U *Crusader* flew. *National Archives*

The outbreak of hostilities in Korea 25 June 1950 created worldwide tension. After just 47 days at home, a new air group, CVG-7, was loaded aboard and *Midway* redeployed to the Mediterranean. Sailing 7 July, *Midway* had her first jet fighter squadron aboard. The F9F-2 *Panther*s assigned to VF-71 could attain 575 mph at sea level. VF-72's F8F-1B *Bearcat*s replaced the dated WWII-era F4Us for this deployment. The F8F was rated at 434 mph. This was *Midway*'s first deployment with jet aircraft and her only deployment with Grumman's powerful piston-engine *Bearcat* fighter. It was also the end of the line for the WWII-era TBM *Avenger*s aboard *Midway*. She arrived at Gibraltar, 20 July 1950. Above: A pair of TBM-3Rs are readied for deck launch over the stern. *National Archives*

CVG-7
10 July 1950 – 10 November 1950
Mediterranean
CO - Captain Frederick N. Kivette
CAG - CDR Henry E. McNeely

VF-71	F9F-2	L 1xx
VF-72	F8F-1B	L 2xx
VF-73	F4U-4	L 3xx
VMF-211	F4U-4	AF 4xx
VA-75	AD-4	L 5xx
VC-4 Det 5	F4U-5N	NA xx
VC-12 Det 5	TBM-3E	NE xx
VC-62 Det 5	F4U-5P	PL xx
HU-2 Det 5	HO3S-1	UR xx

Commander Air Group Seven
CDR Henry E. McNeely

CVG-6
9 January 1952 – 5 May 1952
Mediterranean
CO - Captain Kenneth Craig
CO - Captain Frank O'Beirne
CAG - CDR Ernest W. Humphrey

VF-21	F9F-2	C 1xx
VF-61	F9F-2	C 2xx
VF-41	F4U-4	C 3xx
VMF-225	F4U-4	WI xx
VA-25	AD-4/4L	C 5xx
VC-4 Det 5	F4U-5NL	NA xx
VC-12 Det 5	AD-4W	NE xx
VC-33 Det 5	AD-4NL/4Q	SS xx
VC-62 Det 5	F2H-2P	PL 9xx
HU-2 Det 5	HUP-1	UR xx

Commander Air Group Six
CDR Ernest W. Humphrey

The Piasecki HUP *Retriever* was a single engine tandem rotor utility helicopter. HUPs operated from *Midway* from 1952 through 1962. The prototype was designated the XHJP-1, and first flew in March 1948. It entered service in 1949. The design utilized two, three bladed 35 foot diameter rotors, which could be folded for storage. HUPs were powered by a single 550 hp Continental R975-46 radial engine. Maximum speed was 108 mph and it had a range of 360 miles. A total of 339 aircraft were delivered over the 15 year service life of the aircraft.

The HUP was produced for the Navy in four versions. The HUP-2 was the first production helicopter equipped with an auto-pilot. To provide rescue without crew assistance, an electrically operated door, available after folding the copilot's seat forward, opened through which a rescue sling could be lowered from an overhead winch. The last HUPs were withdrawn from service in 1964. The HUP had a crew of two and could carry four passengers. It was 32 feet long and weighed 4,100 lb empty. It's maximum load was 1,650 lb. The service ceiling was 10,200 ft. *National Archives*

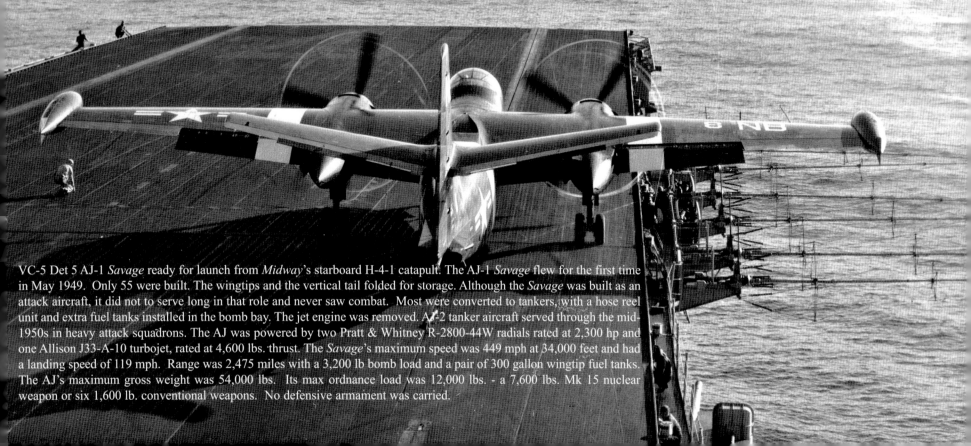

VC-5 Det 5 AJ-1 *Savage* ready for launch from *Midway*'s starboard H-4-1 catapult. The AJ-1 *Savage* flew for the first time in May 1949. Only 55 were built. The wingtips and the vertical tail folded for storage. Although the *Savage* was built as an attack aircraft, it did not to serve long in that role and never saw combat. Most were converted to tankers, with a hose reel unit and extra fuel tanks installed in the bomb bay. The jet engine was removed. AJ-2 tanker aircraft served through the mid-1950s in heavy attack squadrons. The AJ was powered by two Pratt & Whitney R-2800-44W radials rated at 2,300 hp and one Allison J33-A-10 turbojet, rated at 4,600 lbs. thrust. The *Savage*'s maximum speed was 449 mph at 34,000 feet and had a landing speed of 119 mph. Range was 2,475 miles with a 3,200 lb bomb load and a pair of 300 gallon wingtip fuel tanks. The AJ's maximum gross weight was 54,000 lbs. Its max ordnance load was 12,000 lbs. - a 7,600 lbs. Mk 15 nuclear weapon or six 1,600 lb. conventional weapons. No defensive armament was carried.

In a spectacular accident occurring 11 November 1951, LT Earl W. Keegan, Jr in a VF-21 *Panther* missed all the arresting wires and hit the barricade. The aircraft engaged the barricade but broke free and continued forward, impacting a pack of four F9Fs parked on the port bow. Keegan's disintegrating F9 and another *Panther* went over the port catwalk into the water inverted. The collision started a large fire in the wreckage of four F9s on the flight deck. One plane captain was killed and two fueling crewmembers were badly burned. LT Keegan survived, despite the crash impact and landing in the water inverted. He later commanded VU-10 at Guantanamo in F-8 Crusaders and retired as a CDR in 1969. *National Archives*

A F4U-5NL *Corsair* assigned to VC-4 deck launches 28 August 1952. The Vought F4U-5 was a high-altitude fighter, designed to operate up to 45,000 feet. It was fitted with a supercharged Pratt and Whitney R-2800-32W engine, developing 2,300 hp, 200 more than the F4U-4. Maximum speed was 469 miles per hour at 26,800 feet and rate-of-climb was 3,780 feet/min at sea level. The entire outer-wing panels, for the first time on a *Corsair*, were metal covered. The nose was dropped about 2 degrees to improve longitudinal stability and vision. Production began in 1946 with an order for 223 aircraft.

The requirement for night and all-weather fighters had grown to such an extent that the Navy ordered a large number of aircraft in the first block built as night fighters designated F4U-5N's. This version is easily distinguished by its two-foot diameter radar dome in the leading edge of the starboard wing. The F4U-5NL was an all-weather version of the F4U-5N. It was essentially identical to the F4U-5 except that it included provisions for both night-fighter and cold weather operations which installed de-ice boots on the wings and empennage, and de-ice shoes on the propeller blades. USN

A section of McDonnell F2H-2B *Banshees* over the Mediterranean, 28 March 1953. They are assigned to VC-4 Det 5 aboard *Midway*. The F2H-2 version of the *Banshee* differed from the F2H-1 by being fitted with 200 gallon wingtip tanks. It also had external racks which could carry two 500 pound bombs or six 5-inch rockets. The fuselage of the F2H-2 was slightly longer and larger capacity fuel tanks were fitted. The F2H-2 was ordered in May 1948, and the first aircraft (BuNo 123204) flew on 18 August 1949. A total of 364 F2H-2 *Banshees* were built, the first being delivered in 1949 and the last in May 1952. The F2H-2B was a nuclear strike version of the F2H-2. It was externally similar to the F2H-2, but had local strengthening of the wings to carry a 1,650 lb Mk 7 or a 3,230 lb Mk 8 nuclear weapon under the port wing.

The F2H-2B was manufactured alongside the basic F2H-2 on the McDonnell production line. The F2H-2 was powered by twin Westinghouse J34-WE-34 turbojets of 3,250 lbs thrust each and could reach 575 mph at sea level. Its initial rate of climb was 7,300 feet per minute and it's service ceiling 48,500 feet. The landing speed was 102 mph. Normal range was 1,200 miles and combat range was 620 miles. Fuel capacity was 877 gallons. With the two wingtip tanks fitted, a total of 1,277 gallons could be carried. The F2H weighed 11,146 lbs. empty and 21,000 lbs. at maximum takeoff. Armament consisted of four 20mm M3 cannons with 150 rounds per gun and a maximum external ordnance load of 1,000 lbs. could be carried. A typical load consisted of two 500 lb. bombs or six 5-inch rockets. *National Archives*

A VC-62 Det 35 F2H-2P *Banshee* over the Mediterranean during *Midway's* 1954 deployment. The F2H-2P was an unarmed photographic reconnaissance version of the F2H-2. It had the distinction of being the first jet-powered reconnaissance aircraft built for the Navy. It differed from the standard F2H-2 in having a widened and longer nose to provide space for six vertical and oblique cameras. Total length was increased to 42 feet 5 inches. All armament was removed. For night photography, a container for 20 flash cartridges could be carried underneath each wing. The first F2H-2P (BuNo 123366) was started as the 184th F2H-2, but was modified during production. It flew for the first time 12 October 1950. A total of 89 F2H-2Ps were built, the last being delivered 28 May 1952, the last F2H-2 of any type delivered.
National Archives

A section of Douglas AD-4NL *Skyraiders*, assigned to VC-33 Det 35, returns to *Midway* during her 1954 Mediterranean deployment. The AD-4N was a three-seat night attack version of the AD-4. A searchlight was carried on the port underwing pylon, and a radar was fitted on the starboard underwing pylon. A total of 307 AD-4Ns were built. AD-4N BuNo 124153 was modified at Douglas with anti-icing and deicing equipment, and it's armament increased to four 20mm cannons. 36 additional AD-4Ns were thus reconfigured in service. The 37 aircraft with these modifications completed were redesignated AD-4NLs. The AD-4N was powered by a Wright R-3350-26WA Duplex *Cyclone* 18-cylinder radial engine rated at 2,700 hp at takeoff and 2,100 hp at 14,500 feet. It gave the aircraft a maximum speed of 350 mph at 18,300 feet. The AD-4NLs maximum range was 1,100 nautical miles. Its empty weight was 11,400 lbs. and max gross weight was 24,000 lbs. *National Archives*

VF-31 McDonnell F2H-3 *Banshees* return to *Midway* in the Mediterranean, 16 June 1954. The F2H-3 was an all-weather fighter adaptation of the basic *Banshee* design. It had a Westinghouse APQ-41 radar installation housed inside the nose of an 8-foot longer fuselage. The horizontal stabilizer on the F2H-2 had been mounted on the vertical tail, but on the F2H-3 it was moved downward to the rear fuselage tailcone. The last of 250 F2H-3s was delivered to the Navy 31 October 1953. The F2H-3 became the Navy's standard carrier-based all-weather fighter for much of the 1950s. Its maximum speed was 524 mph at 35,000 feet with an initial climb rate 5,900 feet per minute. The F2H-3 landed at 114 mph and had a service ceiling was 46,600 feet. Combat range was 620 miles and maximum takeoff weight was 28,500 lbs. Armament consisted of four 20mm cannons. Four weapons racks were provided for bombs of up to 500 lbs. or HVAR/HPAG rockets. *National Archives*

CVG-6
4 January 1954 – 4 August 1954
Mediterranean
CO - Captain Clifford S. Cooper
CO - Captain William H. Ashford, Jr.
CAG - CDR Carlton H. Clark

VF-31	F2H-3	K 1xx
VF-33	F9F-6	K 2xx
VF-73	F9F-6	L 3xx
VA-25	AD-4B/AD-6	C 5xx
VC-12 Det 35	AD-4W	NE 7xx
VC-33 Det 35	AD-3Q/4NL	SS 8xx
VC-62 Det 35	F2H-2P	PL 9xx
HU-2 Det 35	HUP-2	UR xx
VC-5 Det 35	AJ-1	NB xx

Commander Air Group Six
CDR Carlton H. Clark

VF-73 F9F-6 *Cougars* overfly *Midway* in the Mediterranean, 16 June 1954. The XF9F-2/ XF9F-3 *Panther* contract awarded to Gruman in October 1946 included a clause for design data on a swept-wing version. The development of a swept-wing *Panther* was made more urgent by the appearance of the MiG-15 in Korea in November 1950. The MiG-15 was powered by a derivative of the same Rolls-Royce *Nene* as was the *Panther*, but was nearly 100 mph faster. The Navy and Grumman agreed that it was urgent to accelerate the development of a swept-wing version of the *Panther*. A contract for the modification of three F9F-5 airframes was signed 2 March 1951. Grumman's Design 93 was a swept-wing conversion version of the *Panther*. It retained the fuselage, vertical stab, engine, and undercarriage of the F9F-5, but was fitted with wings swept at 35 degrees and swept horizontal tail surfaces. In order to reduce the approach and stalling speeds to acceptable levels, the chord of the leading-edge slats and the trailing-edge flaps were both increased. Much larger split flaps were fitted underneath the fuselage center section. The fuselage was increased in length by two feet, the the wingroot-mounted intakes were extended forward and the wing root fillets were enlarged. The internal fuel capacity was only 919 US gallons, as compared with 1,003 US gallons for the F9F-5. The F9F-6 *Cougar* flew the first time 20 September 1951. Large wing fences were found necessary to inhibit spanwise airflow and to preserve lateral control effectiveness. The prototype F9F-6 actually had better carrier handling characteristics than the straight-winged F9F-5. The first 30 production F9F-6s were powered by the J48-P-6A turbojet with 7,000 lbs. thrust, but the remainder of F9F-6 production was fitted with the 7,250 lbs. thrust J48-P-8. The armament consisted of four 20mm cannon and two wing racks for 1,000 lb. bombs or 150 US gallon drop tanks. The first squadron to receive the F9F-6 was VF-32, which transitioned to the *Cougar* in November 1952. The *Cougar* arrived too late to fly combat sorties in Korea. The last of 646 F9F-6 *Cougars* was delivered 2 July 1954. *National Archives*

16

Commander Air Group One
CDR William B. Morton

CVG-1
27 December 1954 – 14 July 1955
World/WestPac
CO - Captain Reynold D. Hogle
CAG - CDR William B. Morton

VF-101	F2H-2/F2H-2B	T 1xx
VF-12	F2H-2	T 2xx
VF-174	F9F-6	R 4xx
VA-15	AD-6	T 5xx
VC-4 Det 35	F2H-4	T 6xx
VC-12 Det 35	AD-4W	T 7xx
VC-33 Det 35	AD-5N	T 8xx
VC-62 Det 35	F2H-2P	T 9xx
HU-2 Det 35	HUP-2	UR xx

Ending *Midway*'s long association with CVG-6, *Air Group One* was embarked for the 1954 work-up cycle. The composition of the air group was dominated with *Banshees*. After just four-and-a-half months at home, *Midway* and CVG-1 departed Norfolk 27 December 1954 on a world cruise that would cover 22,000 miles. It was also her last as a straight-deck carrier. After rounding the Cape of Good Hope, *Midway* transferred to Pacific Fleet. She would remain there for the duration of her commissioned service. The ship was drydocked during a port visit to Yokosuka, from 21 through 26 May 1955.

WORLD CRUISE AND THE PACIFIC FLEET 1955

VA-15 AD-6 Skyraiders on *Midway's* bow while drydocked in Yokosuka, Japan, 23 May 1955. *National Archives*

Arriving at her new homeport of Alameda in 1958, *Midway* began work-up exercises with *Air Group Two*. Her marriage with CVG-2 would last through the next eight years and six 7th Fleet deployments. An "NE" tail-code identified an aircraft as assigned to CVG-2/*Midway* from 1958 through 1965. *Air Group Two* was assigned two supersonic jet fighter squadrons. VF-64 flew the F3H-2 *Demon*, capable of firing the radar-guided *Sparrow* air-to-air missile. The *Demon* was rated at 636 mph at sea level. VF-211 was equipped with the record-setting F8U-1 *Crusader*. Although a difficult aircraft to bring aboard, the F8U-1 could easily exceed 1,100 mph. It was the first ever deployment of the reconnaissance variant F8U-1P, assigned to VFP-61. VAH-8 with 10 enormous nuclear-capable *Skywarriors*, made six consecutive A3D deployments in *Midway*.

**Commander
Carrier Air Group Two**
CDR John R. Bowen

CVG-2
16 August 1958 – 12 March 1959
WestPac
CO - Captain John T. Blackburn
CAG - CDR John R. Bowen

Squadron	Aircraft	Code
VF-211	F8U-1	NE 1xx
VF-64	F3H-2	NE 2xx
VA-63	FJ-4B	NE 3xx
VAH-8	A3D-2	ZD 4xx
VA-65	AD-6	NE 5xx
VAW-11 Det Alfa	AD-5W	RR 7xx
VAAW-35 Det Alfa	AD-5N	VV 8xx
VFP-61 Det Alfa	F8U-1P	PP 9xx
HU-1 Det 1/Unit 13	HUP-2	UP xx
CAG	AD-5	NE xx

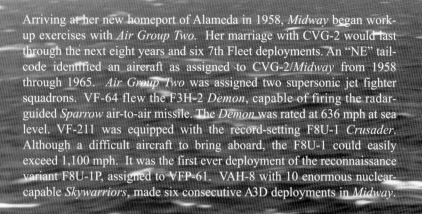

VA-63 *Fury* bolters during air operations in the Western Pacific, 13 October 1958. The FJ-4B was an attack version of the FJ-4. FJ-4B wings were strengthened to fit six underwing ordnance stations capable of carrying a total of 6,000 lbs. of stores. The FJ-4B could carry a nuclear weapon on the port mid-wing station. The last of 222 FJ-4Bs were delivered in May 1958. It was powered by a Wright J65-W-16A turbojet rated at 7,700 lbs. thrust which enabled 680 mph at sea level and an initial climb rate of 7,660 feet per minute. Four internal 20mm cannons were carried. USN

Midway departing Alameda, 15 August 1958. Six different aircraft types are visible in this pack of *Air Group Two* aircraft as *Midway* passes under the Golden Gate Bridge en route to Westpac. *William T. Swisher*

A VF-211 F8U-1 *Crusader* traps aboard *Midway*, c. December 1958. The *Crusader* character-istically would rise well up off of it's main mounts during the arresting gear run out. The Mk 7 Mod 1 arresting gear in *Midway* was operating near the upper end of it's capacity in handling the F-8 and A-3 aircraft. Later the F-4 *Phantom* would tax it to it's technical limit. USN

The F8U-1 was powered by a Pratt & Whitney J57-P-4A/-12 turbojet producing 10,000 lbs. thrust and 16,200 lbs. thrust in afterburner. On 21 August 1958 CDR Duke Windsor in a F8U-1 attained an average speed of 1,015.428 mph on a 15 km course at an altitude of 40,000 feet over China Lake, California. This set a new national speed record, and for this feat the Thompson Trophy was awarded to the Navy and to Vought. The F-8's maximum speed at sea level was 733 mph. The stall speed was a very high 155 mph making the *Crusader* a difficult aircraft to recover on an aircraft carrier.

The *Crusader*'s initial rate of climb rate exceeded 20,000 feet per minute. Its combat radius was 389 miles with an internal fuel capacity of 1,275 gallons. The F8U-1 weighed 15,513 lbs. empty and 27,468 lbs. at maximum takeoff. It was armed with four 20mm Colt-Browning Mk 12 cannons with 144 rounds per gun. Two AIM-9A *Sidewinder* mis-siles could be carried on the fuselage rails. A rocket pack carrying 32 2.75-inch folding-fin rockets could be fitted in the fuselage and lowered with the speed brake. USN

A section of VF-64 F3H-2 *Demons* over Mount Fujiyama, c. December 1958. In November 1952, before the first production J40 powered F3H-1N had flown, McDonnell recommended using the J71 engine in the *Demon*. The J71-powered *Demon* was designated the F3H-2N. The first production F3H-2N flew in June 1955. A total of 140 production aircraft which had been ordered as F3H-1Ns, were delivered as F3H-2Ns with J71 engines. The J71-A-2 was rated at 10,000 lbs. thrust and 14,400 lbs. in afterburner. The F3H-2N carried an APG-51 airborne intercept radar set. The armament consisted of four 20mm cannon and AIM-9 *Sidewinder* air-to-air missiles. The *Demon* was optimized as a strike fighter and could carry up to 6,000 lbs. of ordnance (including nuclear) on two fuselage and six wing stations. USN

The Douglas A3D *Skywarrior* was the largest and heaviest aircraft ever designed for use from an aircraft carrier. The fleet version of the *Skywarrior* was the A3D-2. It was first delivered to VAH-2 in 1957. The A3D-2 was powered by two Pratt & Whitney J57-P-10 non-afterburning turbojets, each rated at 10,500 lbs. thrust. Maximum speed was 640 mph at 2,500 feet and 585 mph at 35,600 feet. The A3D-2 began to reach fleet squadrons in 1957. The A3D-2 set many impressive records. An A3D-2 set a US transcontinental speed record, 21 March 1957, with a time of 5 hours 12 minutes. On 6 June a pair of *Skywarriors* took off from *Bon Homme Richard,* steaming off the California coast, flew across the country and landed 4 hours later on *Saratoga* operating off the east coast of Florida. The last of 283 *Skywarriors* rolled off the production line in January 1961. USN

**Commander
Carrier Air Group Two**
CDR Edward B. Hotley Jr.

CVG-2
15 August 1959 – 25 March 1960
WestPac
CO - Captain James H. Mini
CAG - CDR Edward B. Hotley Jr.

VF-24	F8U-1	NE 1xx
VF-21	F3H-2	NE 2xx
VA-22	FJ-4B	NE 3xx
VAH-8	A3D-2	ZD 4xx
VA-25	AD-7	NE 5xx
VA-23	FJ-4B	NE 6xx
VAW-11 Det A	AD-5W	RR 7xx
VCP-63 Det A	F8U-1P	PP 9xx
HU-1 Det 1/Unit A	HUP-2	UP xx

**Commander
Carrier Air Group Two**
CDR Robert J. Selmer

CVG-2
15 February 1961 – 28 September 1961
WestPac
CO - Captain Ralph W. Cousins
CO - Captain Robert G. Dose
CAG - CDR Robert J. Selmer

VF-24	F8U-2	NE 1xx
VF-21	F3H-2	NE 2xx
VA-22	A4D-2	NE 3xx
VA-23	A4D-2	NE 4xx
VA-25	AD-7	NE 5xx
VAH-8	A3D-2	NE 6xx
VAW-11 Det Alfa	WF-2	RR 7xx
VMA-311	A4D-2	WL 8xx
VCP-63 Det Alfa	F8U-1P	PP 9xx
HU-1 Det 1/Unit Alfa	HUP-3	UP xx

Air Group Commander's AD5, assigned to VA-65, leaves the end of *Midway's* angle deck, 18 September 1958. USN

Midway pitches into a long Pacific swell as a VF-21 *Freelancers* F3H-2 is launched. This type of ship motion was characteristic of *Midway* and her sisters for their entire service lives. The J71-powered F3H-2 *Demon* was generally well liked by its pilots. They liked its power-actuated slats, and especially appreciated its handling at high altitude as well as its stability during carrier landings. However, the *Demon* had short legs and relatively low endurance. The last of 239 *Demons* were delivered 8 April 1960. Total *Demon* production was 519 aircraft serving in 22 Navy squadrons. After five years as a first-line fighter, *Demons* were replaced by F-8 *Crusaders* and F-4 *Phantoms*. VF-161 was the last Navy squadron to replace its F-3Bs with F-4Bs in September 1964. USN

Commander
Attack Carrier Air Wing Two
CDR Billy D. Holder

CVG-2
6 April 1962 – 20 October 1962
WestPac
CO - Captain Robert G. Dose
CO - Captain Roy M. Isaman
CAG - CDR Billy D. Holder

VF-24	F8U-2	NE 1xx
VF-21	F3H-2	NE 2xx
VA-22	A4D-2N	NE 3xx
VA-23	A4D-2	NE 4xx
VA-25	AD-6/7	NE 5xx
VAH-8	A3D-2	NE 60x
VAW-11 Det Alfa	WF-2	RR 76x
VAW-13 Det Alfa	AD-5Q	VR 7xx
VFP-63 Det Alfa	F8U-1P	NE 9xx
HU-1 Det 1/Unit Alfa	HUP-3	UP xx

Commander
Attack Carrier Air Wing Two
CDR H. Spencer Matthews Jr

Commander
Attack Carrier Air Wing Two
CDR Charles F. Demmler

CVW-2
8 November 1963 – 26 May 1964
WestPac
CO - Captain Leroy E. Harris
CO - Captain Whitney Wright
CAG - CDR H. Spencer Matthews Jr
CAG - CDR Charles F. Demmler

VF-21	F-4B	NE 1xx
VA-22	A-4C	NE 2xx
VA-23	A-4E	NE 3xx
VF-24	F-8C	NE 4xx
VA-25	A-1J	NE 5xx
VAH-8	A-3B	NE 6xx
VAW-11 Det Alfa	E-1B	NE 7xx
VFP-63 Det Alfa	RF-8A	NE 9xx
HU-1 Det 1/Unit Alfa	UH-2A	UP xx

A VF-21 F-4B *Phantom II* is launched from *Midway's* starboard catapult while the ship is moored at Yokosuka, Japan, c. January 1964. The F-4 has been launched in military power, not afterburner, and likely has only minimum fuel aboard. The C-11-1 catapult was rated at 39,000 lbs at 156 mph, enough to comfortably launch this partially fueled *Phantom*. USN

Air Wing Two Vietnam Combat Operations - 1965

Midway and CVW-2 left Alameda 6 March 1965 en route Southeast Asia. She proceeded to *Yankee Station* and joined Task Force 77. Arriving on station 10 April, *Midway* aircraft commenced combat air operations for the first time in her 20-year history. Strikes against military and logistics installations in North and South Vietnam were carried out in cycles of 30-day line periods broken by 10-day inport repair and replenishment port calls. During the nine-month deployment twenty-two aircraft were lost - ten *Skyhawks*, six *Crusaders*, four *Skyraiders* and two *Phantoms*. Of those, 17 were lost in combat, 12 of which were to AAA or small arms fire. Eleven aircrew were killed in action and five others became POWs. Five operational aircraft losses occurred, but only one pilot was lost. It was a long, hard, demanding cruise. For their performance on this cruise, *Midway* and CVW-2 were awarded a Navy Unit Commendation Medal. *Midway* also earned the *ComNavAirPac* Battle Efficiency "E," recognizing her as the outstanding carrier in the Pacific Fleet. USN

View of the hangar deck while rearming from an ammunition ship alongside, during the 1965 combat deployment. The photograph was taken from the portside of hangar bay 2 looking forward and to starboard into hangar bay 1. The *Skyhawks* belong to VA-22. On the port side is a *Skyhawk* with an NM tailcode from CVW-19 in *Bon Homme Richard* (CVA-31). USN

Secretary of the Navy Paul Nietze who had attended the strike brief prior to launch, congratulates CDR Lou Page on his MiG kill.

On 17 June 1965, *Midway* fighters engaged four MiG-17 *Frescos*. VF-21 XO, CDR Lou Page and his RIO, LT John C. Smith, in an F-4B, scored the first confirmed MiG kill of the Vietnam War when they shot down one of the MiGs. The kill came during a strike against the Than Hoa Bridge, located 80 miles south of Hanoi on a major supply route from Haiphong to the Ho Chi Minh trail. Late in the mission, after the strike group had gone "feet wet," Page and his wingman, LT Dave Batson made a sweep north searching with the radar for any contacts. A pattern of MiGs showing up late in a strike had been noticed. Two blips painted at about 45 miles north of the pair of *Phantoms*. Executing their plan as briefed, Batson fell back into a three-mile trail and waited for

Page to make a positive ID. The plan was to make head-on *Sparrow* shots. Both F-4s accelerated to 500 knots to gain energy. Sighting the F-4s, the MiGs began a turn that revealed them to be MiG-17s. At close to minimum range, Page fired a *Sparrow* at one of the more distant MiGs in the formation, as planned. The missile guided and detonated, causing the MiG to lose a wing and roll out of control. In the trail F-4, Dave Batson's RIO, LCDR Robert Doremus, obtained a lock on the closest MiG, as had been agreed between the RIOs. The *Sparrow* guided, destroying another MiG-17 in the same formation. Both F-4s returned to *Midway* without tanking, landing with only 400 lbs of fuel, not enough for even one more pass. USN

VA-25 A-1H *Skyraider* armed for strike into South Vietnam, c. June 1965.

Right: VA-25 ready room following the MiG-17 kill, 20 June 1965. From left to right: LTJG Jim Lynne (#2 Greathouse wingman), LCDR Ed Greathouse (#1 Flight leader), LT Clint Johnson (#3 2nd section leader), LTJG Charlie Hartman (#4 Johnson wingman). LT Johnson in NE 577 and LTjg Hartman in NE 573 (photograph below) each received a half-credit for the kill. USN

Above: This A-1H *Skyraider* and a sister *Spad* destroyed a MiG-17 with 20mm cannon fire on 20 June 1965. As part of a SAR effort, four A-1H *Skyraiders* from VA-25 had been launched on a mission to locate a downed Air Force photo-recon pilot. The *Skyraiders* were carrying a standard RESCAP load of two 150 gallon drop-tanks, four pods with 19 2.75 inch rockets each and 800 rounds of 20mm. A radar-picket ship operating in the Tonkin Gulf detected two enemy aircraft coming from the north and warned the *Skyraiders*. In what should have been a one-sided fight, two MiG-17s jumped the four *Spads*. The *Skyraiders* immediately jettisoned their ordnance and fuel tanks while diving for the deck. Encountering a small mountain, they started circling it, using it for cover. Two MiG-17s came down and made a pass at the lead *Skyraider*. Coming around the hill LT Clint Johnson saw LCDR Ed Greathouse and LTjg Jim Lynne low with the MiG lined up behind them. He fired a burst and missed, but got the MiG's attention. The MiG turned hard into Johnson to make a head-on pass. Johnson and LTjg Charlie Hartman fired as the MiG passed so close that Hartman thought that his vertical stabilizer had hit his wingman's tailhook. Hartman's rounds appeared to go down the MiG's intake and into the wing root and Johnson's along the top of the fuselage and through the canopy. The MiG did not return fire, rolled inverted, hit a hill, and exploded in a farm field.

30

Commander
Attack Carrier Air Wing Two
CDR Robert E. Moore

CVW-2
6 March 1965 – 23 November 1965
WestPac/Vietnam
CO - Captain James M. O'Brien
CAG - CDR Robert E. Moore

VF-21	F-4B	NE 1xx
VA-22	A-4C	NE 2xx
VA-23	A-4E	NE 3xx
VF-111	F-8D	NE 4xx
VA-25	A-1H	NE 5xx
VAH-8	A-3B	NE 6xx
VAW-11 Det Alfa	E-1B	RR 7xx
VAW-13 Det 1/Unit Alfa	EA-1F	VR xx
VFP-63 Det Alfa	RF-8A	NE 9xx
HU-1 Det 1/Unit Alfa	UH-2A	UP xx

Left: VA-25 *Spad* drops 12 Mk 82 500 lb. bombs on a target in South Vietnam, c. June 1965. USN

A VAH-8 A-3B Skywarrior drops iron bombs on a Viet Cong target, c. July 1965. VAH-8 was assigned 10 aircraft and performed both bombing and inflight refueling missions. USN

The A3D-2 *Skywarrior* was redesignated A-3B in 1962. The first 123 A3D-2s differed from the A3D-1 by having a strengthened airframe and in being fitted with more powerful engines. The bomb bay was redesigned to carry a wider range of internal stores. The next 20 A3D-2s were fitted with a inflight refuelling probe on the port side that extended forward of the nose. Provision was made for the installation of a removable inflight refuelling and tanker package in the bomb bay, which included a hose reel unit and a large fuel tank. The last 21 A3D-2s had an ASB-7 bomb director system installed beneath a flat panel nose radome which replaced the original pointed radome.

The shape of the A3D-2 tail was markedly altered. The tail turret was eliminated, and was replaced by a set of electronic countermeasures equipment mounted inside a new dovetail-shaped fairing. The inflight refuelling package, the flat panel nose radome, and the new dovetail fairing were retrofitted to most early A3D-2s. *Skywarriors* were assigned to 13 VAH squadrons. The fuselage shape and it's size earned the aircraft the nickname of *Whale*. It's combat radius was 1,325 miles with a 4,100 lb. offensive load. The A3D-2 weighed 39,409 lbs. empty and 82,000 lbs. at maximum takeoff. A maximum weapons load of 12,000 lbs. could be carried in the internal bomb bay.

VA-23 A-4E *Skyhawk* spotted on the port angle during rearming operations c. July 1965.

The *Skyhawk* was one of the primary Navy attack aircraft of the Vietnam War. *Skyhawks* were often attacked by MiGs, but their maneuverability made it possible for them to evade most of these attacks. Although the A-4 was not designed as a fighter, it did achieve an air-to-air kill. On 1 May 1967, LCdr Ted Swartz of VA-76 flying from *Bon Homme Richard* shot down a MiG-17 with air-to-ground rockets. This was the only A-4 air-to-air kill of the Vietnam War. Losses of *Skyhawks* to AAA and SAMs were heavy, especially during the early stages of the war. Initially, attacks were flown at low level to avoid detection by radar, but it was found that the required climb before bomb release resulted in an unacceptable loss of speed and maneuverability. A change of tactics was made. Targets were approached at high altitude at high speed and were bombed in shallow diving attacks. A high-altitude approach brought vulnerability to surface-to-air missiles, but North Vietnamese SA-2 SAMs could be avoided, provided that the pilot saw one coming. *Skyhawks* were armed with *Shrike* anti-radiation missiles to suppress AAA and SAM sites.

The A-4E was a considerably advanced version of the *Skyhawk*. It had a new engine, the Pratt & Whitney J52-P-6A, rated at 8,500 lb.st. This engine yielded 27 percent lower fuel consumption than the Wright J65 of earlier versions. The J52 engine required a redesign of the center fuselage and intake ducting, adding a splitter plate next to the fuselage. The A4D-5 optimized the basic *Skyhawk* design to deliver conventional munitions adding an extra pair of underwing weapons hardpoints. These extra stations increased the maximum weapons load to 8,200 lbs. The nose was lengthened by 14 inches to accommodate an ASN-19A navigation computer. The A-4E began to replace the A-4B in December 1962 with VA-23 at Lemoore. A total of 500 A-4Es were built. USN

VAW-11 Det Alfa "*Willy Fudd*" over the ramp during recovery c. June 1965. In 1955, BuAer issued a request for an advanced carrier-based AEW aircraft, which was ultimately the E-2 *Hawkeye*. However, BuAer realized that the spec was so advanced that the new aircraft couldn't be ready until the 1960s, and so also ordered two prototypes of a much simpler AEW derivative of the proven S2F *Tracker* antisubmarine warfare aircraft as an interim solution. The result was the WF-2 *Tracer*. The WF-2 entered the fleet in 1960 and was reclassified as the E-1B in October 1962.

The *Willie Fudd* was a capable machine, with a then state-of-the-art APS-82 radar and analogue tactical datalinks, it remained in service until 1977. It was powered by two Wright R-1820-82A radials of 1,525 hp each. Its maximum speed was just 238 mph but it had a range of more 1,000 miles. Its maximum gross weight was 26,600 lbs. "*Willy Fudds*" flew from *Midway* during four deployments from 1961 through 1965. The affectionate unofficial tag "*Willy Fudd*" was derived from the aircraft's original WF-2 designation. USN

VF-111 *Sundowner* F-8D during the 1965 Vietnam combat cruise. The *Crusader* was armed with four 20mm cannons and could carry up to four *Sidewinders* when fitted with the Y-rail station as this aircraft is.   It was powered by an afterburning Pratt & Whitney J57-P-20A turbo-jet rated at 18,000 lbs thrust. At a combat load weight of 24,400 lbs the thrust-to-weight ratio was a respectable 1 to 1.3.

The F-8 could attain Mach 1.72, or more than 1,100 mph. From 1965 to 1968 the *Crusader* was the leading MiG-killer over Vietnam, accounting for a total of 18 confirmed victories. During the war the *Crusader* had the highest kill ratio over communist jets of any Navy aircraft. Overall, the *Crusader* achieved a six-to-one kill-loss ratio. USN

This F-4B, BuNo 151488, was flown by CDR Lou Page and LT John Smith on 17 June 1965 when they shot down a MiG-17 over North Vietnam. CDR Bill Franke's name is on the canopy rail. USN

Left: On 24 August 1965 CDR Bill Franke, the CO of VF-21, was flying from *Midway* in this F-4B, BuNo 152215. On his second mission of that day, a raid on the Tan Hoa bridge, his F-4B was hit by a SAM. He and his RIO ejected seconds before the aircraft exploded. As he was drifting to earth he saw Vietnamese waiting for him below. He had suffered a broken wrist and had shrapnel wounds from the SAM hit and subsequent ejection from the aircraft. Before he ever hit the ground the Vietnamese were on him with machetes. Bill Franke was a POW in Hanoi for 7 1/2 years. He was released in January 1973 and became the CO of VX-4. He retired as a Captain in 1977. USN

36

III - THE NEW *MIDWAY* AND *CARRIER AIR WING FIVE*

CVW-5
16 April 1971 – 6 November 1971
WestPac/Vietnam
CO - Captain William L. Harris, Jr.
CAG - Captain Ralph B. Rutherford

VF-161	F-4B	NF 1xx
VF-151	F-4B	NF 2xx
VA-93	A-7B	NF 3xx
VA-56	A-7B	NF 4xx
VA-115	A-6A	NF 5xx
	KA-6D	NF 52x
VFP-63 Det 3	RF-8G	NF 60x
VAQ-130 Det 2	EKA-3B	NF 610
HC-1 Det 8	SH-3G	NF 00x
VAW-115	E-2B	NF 01x

Commander
Attack Carrier Air Wing Five
Captain Ralph B. Rutherford

VAQ-130 Det 2 *Skywarrior* comes aboard *Midway* in September 1971. In 1967, 85 A-3Bs were modified at the Naval Air Rework Facility (NARF) Alameda into aerial tankers by removing their bombing equipment and installing permanent fuel management and fuel transfer systems. The reel and receptacle for the probe-and-drogue system was installed in the rear part of the bomb bay, with the hose and attached basket being deployed via a protrusion which extended below the rear fuselage. These modified aircraft were redesignated KA-3Bs.

Later, 34 of these KA-3Bs were modified as electronic countermeasures/aerial tanker aircraft. They were redesignated EKA-3Bs. The ECM equipment was installed in the forward portion of the bomb bay, on top of the tail, in the dovetail tail fairing, in an external canoe-shaped fairing beneath the fuselage and in pods attached to both sides of the fuselage fore and aft of the wing. After 1975, most surviving EKA-3Bs had their ECM equipment and external fairings removed, and were redesignated KA-3Bs. USN

Vietnam Combat Operations 1971 - 1972

A section of VF-161 *Chargers* F-4Bs tank from a VAQ-130 Det EKA-3B. *Midway* departed Alameda for her sixteenth overseas deployment 16 April 1971. CV-41 and CVW-5 arrived at Yankee Station off the coast of North Vietnam 30 April. Relieving *Hancock* (CVA-19) 18 May, CVW-5 immediately commenced strike operations. On 11 June, less than a month after commencing combat operations, tragedy struck. A VAW-115 E-2B *Hawkeye* disappeared without a trace while on a ferry flight.

A second E-2 and a VA-56 A-7 were lost 19 October. These were the only three aircraft losses during the deployment. CVW-5 flew more than 6,000 sorties in support of ground forces. Operations were centered on interdicting the Ho Chi Minh Trail. The 4,500 men in *Midway* spent 146 days at sea on this seven-month deployment. The ship was awarded a Meritorious Unit Commendation following her return to Alameda 6 November 1971. USN

CVW-5

10 April 1972 – 3 March 1973
WestPac/Vietnam
CO - Captain S. R. Foley, Jr.
CAG - Captain Carroll E. Myers
CAG - Captain Marion H. Isaacks

VF-161	F-4B	NF 1xx
VF-151	F-4B	NF 2xx
VA-93	A-7B	NF 3xx
VA-56	A-7B	NF 4xx
VA-115	A-6A	NF 5xx
	KA-6D	NF 52x
VFP-63 Det 3	RF-8G	NF 60x
VAQ-130 Det 2	EKA-3B	NF 610
HC-1 Det 2	SH-3G	NF 00x
VAW-115	E-2B	NF 01x

Commander
Attack Carrier Air Wing Five
Captain Carroll E. Myers

VAW-115 E-2B *Hawkeye* on *Midway*'s port C-13 catapult, c. June 1972. The Grumman E-2B was an all-weather command and control aircraft that provided early warning of approaching enemy aircraft as well as the ability to direct interceptors into an attack position. In addition to its primary AEW mission the E-2B also provided strike and traffic control area surveillance, search and rescue coordination, navigation assistance and communications relay. A normal compliment of four E-2s were assigned to VAW-115 in *Midway*. *CDR Rob Anderson, USNR (Ret)*

Midway F-4Bs from VF-161 team with A-7Es from *America* (CVA-66) on a radar bombing mission during *Linebacker II* operations in December 1972 at the close of the Vietnam War. Radar bombing was an Air Force tactic and such taskings were not well received by the Navy. USN

VF-161 *Chargers* F-4B in tension on *Midway's* port catapult during 1972 combat deployment to Vietnam. *CDR Rob Anderson, USNR (Ret)*

Ready Room Six in *Midway*, c. summer of 1972. VF-161 crews shot down five *MiGs* during the 1972 deployment while operating from this compartment.

Below: This F-4B, BuNo 153068, crewed by LT H. A. "Bart" Bartholomay and LT Oran O. Brown, shot down a MiG-19 18 May 1972. The engagement was fought within 4 miles of the North Vietnamese fighter base at Kep. On the same flight their wingmen LT Pat Arwood and LT Jim Bell in *Rock River 105* (BuNo 153915) shot down another MiG-19 near Kep. Both kills were made with *Sidewinders*. *Rock River 110* (below) was lost 14 January 1973 to AAA. *Rock River 105* is preserved at the Naval Aviation Museum at Pensacola, FL. *CDR Rob Anderson, USNR (Ret)*

Switchbox 211, F-4B BuNo 150648 over North Vietnam in June 1972. An all-weather fighter, the F-4B was developed as an interceptor to exploit the radar guided *Sparrow* missile system in the fleet air defense role. In actual service it was used for close air support and interdiction as well as air superiority. The F-4B was powered by two afterburning General Electric J79-8A turbo-jets of 17,000 lbs thrust each. First flown in 1958 it set a more than a dozen speed, altitude and time to climb world records which stood until the F-15 *Eagle* entered service in 1972.

In one of these record flights *Skyburner*, an early F4H-1, set a world record in 1961 of 1,606 mph. The F-4Bs maximum service speed was 1,485 mph at 48,000 feet or 845 mph at sea level. At 38,500 lbs combat weight, the twin J79s yielded a thrust to weight ratio of 1 to 1.1. Of the 57 Navy aerial victories in Vietnam 36 were in *Phantoms*. McDonnell-Douglas built more than 5,000 *Phantoms* including 1,200 for the Navy, 2,600 for the Air Force, and more than 1,000 for Allied nations. USN

A section of *Sidewinder* armed VA-93 *Ravens* A-7Bs overfly *Midway* c. July 1972. The A-7B differed from the A primarily in having the 12,200 lbs. thrust TF30-P-8 engine in place of the 11,350 lb. version in the A. It also had improved flaps. The first A-7B, Bu No 154363, flew on 6 February 1968, test pilot Joe Engle at the controls. The first operational A-7Bs went to VA-146 and VA-215 in late 1968. They deployed aboard *Enterprise* on 6 January 1969. The last of 196 A-7Bs was delivered in May 1969. A-7Bs made 15 combat cruises, losing 11 aircraft in combat and 12 to operational accidents. Seven of the combat losses were to AAA, one to a SAM, and three to undetermined causes.

A-7B squadrons made 45 overseas carrier deployments, the last aboard *John F. Kennedy* in 1977. Remaining A-7Bs were relegated to reserve units until January 1987. The A-7B could reach 683 mph at sea level with an initial climb rate of 7,920 feet per minute. It's landing speed was 139 mph. It weighed 16,100 lbs. empty and 37,000 lbs. at maximum gross. Armament was two 20mm Mk 12 cannon with 600 rounds per gun and up to 15,000 lbs. of ordnance could be carried on eight hardpoints. USN

Japan - 1973
Overseas Family Residence Program

Commander
Attack Carrier Air Wing Five
CDR John B. Streit

CVW-5
11 September 1973 – 5 October 1973
OFRP Yokosuka, Japan
CO - Captain R. J. Schulte
CAG - CDR John B. Streit

VF-161	F-4N	NF 1xx
VF-151	F-4N	NF 2xx
VA-93	A-7A	NF 3xx
VA-56	A-7A	NF 4xx
VA-115	A-6A/B	NF 5xx
	KA-6D	NF 52x
VFP-63 Det 3	RF-8G	NF 60x
(to April 74)		
HC-1 Det 2	SH-3G	NF 00x
VAW-115	E-2B	NF 01x
VQ-1	EA-3B	PR xx

VF-161 F-4N near Mt. Fujiyama in November 1973. For *Midway*'s change of homeport to Yokosuka, Japan, both VF-151 and VF-161 exchanged their weary war-cruise F-4Bs for 24 "Bee-Line" F-4Ns fresh off the production line at NARF North Island. The F-4N was an extensively rebuilt F-4B that included the accomplishment of all outstanding air frame changes, complete rewiring, as well as the installation of the VTAS (Visual Target Acquisition System). This was the first overseas employment of F-4Ns. CDR Rob Anderson, USNR (Ret)

VFP-63 Det 3 RF-8G flying from *Midway*, c. June 1974. The first remanufacture program of the RF-8A reconnaissance version of the *Crusader* rebuilt 73 RF-8As as RF-8Gs. They were returned to Vought for modernization which included the installation of the J57-P-22 engine rated at 10,700 lbs. thrust and 18,000 lbs. in afterburner. The ventral fins of later *Crusader* models were provided as well as improved navigation and electronic equipment. In addition, underwing hardpoints for drop tanks were installed and four cameras were mounted in the fuselage reconnaissance bay. The first group of 53 were reworked from 1965-67 and the second block of 20 were completed in 1968-70. The Gs could readily be distinguished from the As by the presence of the long ventral strakes mounted on the aft fuselage. The first RF-8G reentered service in October 1965. It made its first cruise in July 1966 aboard *Coral Sea*. The service life of the RF-8G proved to be much longer than even the most optimistic projections. In the late 1970s, there were substantial numbers of these aircraft still flying in VFP-63.

A second upgrade of reconnaissance *Crusaders* was carried out commencing in February 1977. Many of the J57-P-22 engines of the RF-8Gs were replaced by more powerful J57-P-429 engines. New electrical wiring was installed and new electronic countermeasures equipment was added. These modified RF-8Gs could be identified by two large afterburner cooling air intakes mounted on their upper tailcones, a feature which had first appeared on the F8U-2 (F-8C). The modified RF-8Gs was also fitted with a round antenna mounted at the rear of the upper vertical stab just above the rudder. This carried a radar warning receiver. The RF-8G was the longest-serving US version of the *Crusader*, serving long after its fighter cousins had been withdrawn. The last active Navy unit to fly the RF-8G was VFP-63 Det 2 in *Coral Sea*, which was disestablished in June 1982. The RF-8G flew on with the Naval Air Reserve. VFP-206 and VFP-306, both based at NAF Washington DC, at Andrews AFB, flew RF-8Gs until 29 March 1987. The last *Crusader* in USN service, RF-8G BuNo 146860, was transferred to the Smithsonian Institution the following day. USN

Below: VA-56 *Champs* A-7A c. June 1975. When *Air Wing Five* deployed to Japan in September 1973 they actually reverted from the more powerful A-7B which they had on two prior combat tours. The Ling-Temco-Vought *Corsair II* was to built to replace the A-4 *Skyhawk*. The Navy required low cost and specified that the aircraft be based on an existing design. The requirement did not call for supersonic performance. A maximum bombload of 15,000 lbs. was called for. The A-7 was a shortened version of the F-8 *Crusader* fighter. Because the aircraft did not have to be capable of supersonic performance, it could be shorter, had a wing with less sweep and was powered by a turbofan engine without an afterburner. An important feature of the design was eight external stores positions. The aircraft was armed with a pair of 20mm Mk 12 cannon with 600 rounds per gun, one gun on each side of the intake. The engine used was a Pratt & Whitney TF30-P-6 turbofan delivering 11,350 lbs. thrust. LTV received a contract 19 March 1964 for 7 A-7A flight test articles and 35 A-7A production aircraft. 140 more were ordered 10 November 1965.

The first YA-7A, BuNo 152580, was rolled out on 13 August 1965. It first flew 27 September 1965, with LTV test pilot John Konrad at the controls. Two fleet replacement squadrons, VA-174 and VA-122, received their first A-7As in September and October 1966 respectively. The first operational A-7A squadron was VA-147, established on 1 February 1967. VA-147 embarked on its first combat cruise aboard *Ranger* on 4 November and it's first flew combat missions on 4 December 1967. In comparison with the A-4 *Skyhawk*, the A-7A was easier to maintain and was much more likely to survive combat damage. In addition, the A-7A had considerably longer range, making it possible to fly missions that the A-4 could not support. However, there were initial problems with ingestion of catapult steam into the intake. A total of 193 A-7As were built. Maximum speed was 680 mph at sea level with an initial climb rate of 5,000 feet per minute. The A-7's landing speed was 139 mph. The *Corsair II* weighe15,105 lbs. empty and 34,500 lbs. at max gross. *Hideki Nagakubo courtesy Bob Lawson*

Commander
Attack Carrier Air Wing Five
Captain W. Lewis Chatham

CVW-5
31 March 1975 - 29 May 1975
Operation Frequent Wind
CO - Captain Larry C. Chambers
CAG - Captain W. Lewis Chatham

VF-161	F-4N	NF 1xx
VF-151	F-4N	NF 2xx
VA-93	A-7A	NF 3xx
VA-56	A-7A	NF 4xx
VA-115	A-6A/B	NF 5xx
	KA-6D	NF 52x
VMCJ-1 Det 101	RF-4B	RM 60x
(from April 74)	EA-6A	RM 61x
HC-1 Det 2	SH-3G	NF 00x
VAW-115	E-2B	NF 01x
VQ-1	EA-3B	PR xx

VA-115 *Arabs* A-6A *Intruder*, c. June 1974. The Grumman A2F (A-6) *Intruder* was the result of a February 1957 NAVY RFP for a replacement for the AD *Skyraider*. The most important feature of the design was an advanced all-weather electronic system, called DIANE - Digital Integrated Attack Navigation Equipment. This system had a pair of antennas in the large radome, one for a Norden APQ-92 search radar, and the other for a Naval Avionics Facility APQ-112 track radar. Also included was an ASN-31 inertial navigation system, a Litton APQ-61 ballistics computer, an APN-141 radar altimeter, and an APN-153 Doppler navigation system. The wing had trailing-edge flaps that extended over almost the entire length of the wing operating as flaperons, so conventional ailerons were not fitted. The first aircraft (BuNo 147864) took off 19 April 1960, with test pilot Robert Smyth at the controls. Very early in the flight test program there were problems with the aft-mounted fuselage speed brakes. These brakes had perforations that were designed to smooth airflow and prevent buffeting of the tail surfaces. The fuselage-mounted speed brakes were gradually replaced by unconventional wingtip-mounted speed brakes. Beginning with the 304th aircraft (BuNo 154170) the fuselage-mounted speed brakes were completely omitted. As expected for a complex system, DIANE was initially unreliable. Delivery to the fleet was delayed by almost a year. Carrier trials began aboard *Enterprise* in December 1962. Initial deliveries to VA-42 at NAS Oceana began in February 1963. The A2F-1 was redesignated A-6A 18 September 1962. A total of 488 A-6As were built before production changed to the A-6E version in December 1970. 19 A-6As were converted to A-6Bs, 12 were converted to A-6Cs, 90 became KA-6D tankers, and 13 were converted to EA-6As. Nearly half of the A-6As, 240 total, were converted to A-6Es. The A-6A was powered by two Pratt & Whitney J52-P-6A non-afterburning turbojets producing 8,500 lbs. thrust each. It's maximum speed was 646 mph at sea level with an initial rate of climb of 6,950 feet/min. It's stall speed was 133 mph. Range was 1,350 miles and weighed 25,300 lbs. empty or 53,700 lbs. maximum. A maximum weapons load of 15,000 lbs. could be carried on external pylons, two under each inner-wing and one on the fuselage centerline. *Intruders* operated from *Midway* for 20 years, from 1971 through 1991. Only seven Intruders, four of which were KA-6Ds, were lost during this period and none during the final eight years. Two of the losses were carrier landing accidents, only one was a combat loss, three were to mechanical problems and one to ground impact. *Hideki Nagakubo courtesy Bob Lawson*

Grumman E-2B *Hawkeye* assigned to VAW-115, in flight, c. June 1975. VAW-115 E-2s flew from *Midway* from 1970 to 1991. The E-2A *Hawkeye* was the first aircraft built from the ground up as an AEW platform. The first flight of a prototype in full configuration was 19 April 1961. The *Hawkeye* was fitted with an advanced General Electric APS-96 radar system, with the antenna installed in a 24 foot diameter rotating radome. The flight crew included a pilot, copilot, and three electronics systems officers, all in a pressurized cabin. The aft section was an airborne Combat Information Center (CIC). The CIC had three consoles, all facing to port. The CIC was run by a *CIC Officer* (CICO), who was the mission commander - the pilot and copilot being responsible only for flying the aircraft. The other two operators were the Radar Operator, who supervised the radar system, and the Air Control Officer, who performed air communications. Each sat at a workstation with a large round CRT in the console. Deliveries of the E-2A began in 1964, with a total of 59 production aircraft delivered. The *Hawkeye* was not an overnight success. The E-2A did not meet specifications and was regarded as so unsatisfactory that production was suspended in early 1965. The E-2A's avionics systems were highly capable when they worked, but they were unreliable to the point of being ineffective. The entire *Hawkeye* inventory was grounded at one time. Grumman and the Navy struggled to implement fixes, the result being the E-2B, which included a modernized avionics system built around a Litton L-304 digital computer. The first E-2B flew 20 February 1969. Forty-nine E-2As were converted to E-2Bs through 1971. The *Hawkeye* radar rotodome could be lowered two feet to provide clearance on the hangar deck. In flight, the rotodome was at a positive angle of incidence to provide enough lift to carry its own weight. The APS-96 radar operated in the UHF band and could observe sea and sky over a radius of 200 miles from an operational altitude of 30,000 feet. The radar had some ability to process sea clutter, and most important it provided automatic target tracking. The APS-96 was a major component of the E-2's *Airborne Tactical Data System* (ATDS), which linked the radar with a computer system, datalinks, and an IFF system. The IFF antenna was mounted on the back of the radar antenna, thus the radar scan alternated with IFF interrogation during each rotation. The E-2B weighed 37,678 lbs and had a max gross weight of 51,529 lbs. The *Hawkeye* was powered by twin Allison T56-A-8/8A turboprops rated at 4,050 shp each. These could propel it at a maximum speed of 375 mph. The T56 is also used on the P-3 *Orion* and the C-130 *Hercules* aircraft. The E-2B had a mission radius of 930 miles. Groundwork for the E-2C was laid by the development of a production APS-120 radar. The APS-120 featured an *airborne moving target indicator* (AMTI) capability that gave it a limited ability to detect moving targets on land. The first production E-2C flew 23 September 1972 and entered fleet service in November 1973. Along with the APS-120 radar, the E-2C included an ESM system to locate radar and radio emitters, the ALR-59 Passive Detection System (PDS). Fitting the PDS meant that the E-2C had a longer nose. The E-2C included Link-4 air-to-air and Link-11 air-to-ground datalinks and an improved APX-72 or -76 IFF system. The E-2C had larger cooling package on top of the fuselage behind the cockpit to handle the increased thermal load of the avionics systems. E-2C's were not assigned to VAW-115 until 1985. USN

VF-161 F-4N *Bicentennial* CAG aircraft on approach to NAF Atsugi, c. June 1976.

Commander
Attack Carrier Air Wing Five
CDR John L. Finley

CVW-5
4 October 1975 – 19 December 1975
WestPac/Indian Ocean
CO - Captain Larry C. Chambers
CAG - Captain W. Lewis Chatham
CAG - CDR John L. Finley

VF-161	F-4N	NF 1xx
VF-151	F-4N	NF 2xx
VA-93	A-7A	NF 3xx
VA-56	A-7A	NF 4xx
VA-115	A-6A/B	NF 5xx
	KA-6D	NF 52x
VAW-115	E-2B	NF 60x
VMFP-3 Det 1	RF-4B	RF 61x
VMAQ-2 Det A	EA-6A	CY 62x
HC-1 Det 2	SH-3G	NF 72x

Commander
Carrier Air Wing Five
Captain William B. Kirkconnell

CVW-5
27 September 1977 – 21 December 1977
WestPac/Indian Ocean
CO - Captain D. L. Felt
CAG - William B. Kirkconnell

VF-161	F-4J	NF 1xx
VF-151	F-4J	NF 2xx
VA-93	A-7E	NF 3xx
VA-56	A-7E	NF 4xx
VA-115	A-6E	NF 5xx
	KA-6D	NF 52x
VAW-115	E-2B	NF 60x
VMFP-3 Det 2	RF-4B	RF 61x
VMAQ-2 Det B	EA-6A	CY 62x
HC-1 Det 2	SH-3G	NF 72x

49

VMAQ-2 EA-6A assigned to *Midway*/CVW-5 approaches MCAS Iwakuni, 13 June 1978. Marine EA-6As were assigned to *Air Wing Five* from April 1974 through 1978. The A2F-1Q (EA-6A) was an electronic warfare version of the A2F-1, initially designed as a replacement for the Douglas F3D-2Q *Skyknight* in the Marine Corps. Two A2F-1s were modified as prototypes for the A2F-1Q. The aircraft was redesignated EA-6A in September 1962. 148618 flew for the first time 26 April 1963. The most significant external change was the presence of a canoe-shaped fin-tip fairing to accommodate a set of antennae for a Bunker-Ramo ALQ-86 receiver/surveillance system. This system included ALQ-41, ALQ-51, and ALQ-55 jamming systems. Their primary mission was to suppress enemy electronic activity during air strikes. In addition, the aircraft could carry up to five jammer pods on the underwing pylons and on the centerline. Chaff dispensers could be substituted for the underwing jammer pods. The aircraft retained a limited all-weather attack capability, although EA-6As were very seldom used for offensive operations. The aircraft could carry and launch the AGM-45 *Shrike* anti-radiation missile, but this was very rarely used operationally.

Only 27 EA-6As were built, 2 prototypes, 10 modified from A-6A airframes, and 15 production aircraft built from scratch as EA-6As. The first operational aircraft were delivered to VMCJ-1 at MCAS Cherry Point on 1 December 1965. EA-6As were first deployed to Southeast Asia in October 1966. The aircraft served in Vietnam as supplements and later replacements for Marine EF-10B (F3D-2Q) *Skyknights*. On the 15 purpose-built EA-6As, the ALQ-86 surveillance receiver system and the ALQ-76 jamming systems replaced earlier systems. The purpose-built EA-6As were assigned to three squadrons VMCJ-1, -2, and -3. The Marines retired their last EA-6A in 1985. Although the Navy never used the EA-6A in carrier based squadrons, Marine detachments did deploy aboard *Forrestal* and *Saratoga* in the Mediterranean in the early 1970s. It was powered by two Pratt & Whitney J52-P-6A turbojets of 8,500 lbs. thrust each. Maximum speed was 646 mph at sea level with an initial rate of climb 6,950 feet/min. The EA-6A's stall speed was 133 mph. Its normal range was 1350 miles. Weight was 27,770 lbs. empty or 54,600 lbs. at max gross. A maximum load of 15,000 lbs. could be carried on four underwing hardpoints and one centerline hardpoint. USN

**Commander
Carrier Air Wing Five**
Captain Stewart D. Langdon

CVW-5
7 April 1979 – 18 June 1979
Indian Ocean/North Arabian Sea
CO - Captain Thomas F. Brown
CAG - Captain Stewart D. Langdon
CAG - Captain Steven R. Briggs

VF-161	F-4J	NF 1xx
VF-151	F-4J	NF 2xx
VA-93	A-7E	NF 3xx
VA-56	A-7E	NF 4xx
VA-115	A-6E	NF 5xx
	KA-6D	NF 52x
VAW-115	E-2B	NF 60x
VMFP-3 Det 2	RF-4B	RF 61x
VMAQ-2 Det X	EA-6B	CY 62x
HC-1 Det 2	SH-3G	NF 72x

Right: CVW-5 aircraft pack the stern while inport
Subic Bay, February 1980. Aircraft from each of
the five tactical squadrons are visible.
CDR Pete Clayton

51

Commencing in May 1977, VF-151 and VF-161 transitioned to F-4Js. The replacement aircraft were transferred from VF-191 and VF-194 upon completion of their 1977 deployment in *Coral Sea*, as CV-43 was departing from WestPac. The F-4J, fitted with the AWG-10 pulse-doppler radar, could "look down" to find enemy aircraft in ground clutter. This, plus its more powerful J79-10 engines, made the F-4J a significant improvement over the F-4N.

VA-56 and VA-93 swapped their aging A-7As for the greatly improved A-7E. The "Echo" model A-7 was powered by an Allison/Rolls Royce TF41-A-400 turbofan engine rated at 15,000 lbs. thrust. A single internally mounted M61A1 20mm six-barrel cannon was fitted. A fully integrated digital navigation/weapon delivery system, based on state-of-the-art electronics and an automation design philosophy provided new mission effectiveness and flexibility. The A-7E was consistently capable of delivering bombs with an accuracy of less than 10 mils Circular Error Probable (CEP). It was common for the A-7E to destroy the 4 ft square practice bombing spar, towed behind the carrier, with direct hits with tiny inert Mk 72 practice bombs.

At the same time, VA-115 exchanged their A-6As and Bs for A-6Es. The A-6E was a quantum leap for the *Intruder* community. The Norden APQ-148 radar replaced the unreliable and maintenance-intensive APQ-92 and -112 units. The new aircraft included a truly digital attack and navigation system and a built-in test function. The uprated J52-P-8B engines delivered 9,300 lbs thrust per engine. *Midway* deployed to Japan four years earlier with arguably an obsolete air wing, but by the close of 1977 carried a far more combat-capable air wing.

VMAQ-2 Det Y EA-6B on approach to Atsugi, c. July 1979.

Commander
Carrier Air Wing Five
Captain Steven R. Briggs

CVW-5
30 September 1979 – 20 February 1980
Indian Ocean/North Arabian Sea
CO - Captain E. Inman Carmichael
CAG - Captain Steven R. Briggs

VF-161	F-4J	NF 1xx
VF-151	F-4J	NF 2xx
VA-93	A-7E	NF 3xx
VA-56	A-7E	NF 4xx
VA-115	A-6E	NF 5xx
	KA-6D	NF 52x
VAW-115	E-2B	NF 60x
VMFP-3 Det 3	RF-4B	RF 61x
VMAQ-2 Det Y	EA-6B	CY 62x
HC-1 Det 2	SH-3G	NF 72x

VAQ-136 EA-6B on approach to Atsugi, c. December 1981.

CVW-5
14 July 1980 – 26 November 1980
Indian Ocean/North Arabian Sea
CO - Captain E. Inman Carmichael
CAG - Captain Steven R. Briggs
CAG - Captain Roger P. Flower

VF-161	F-4J	NF 1xx
VF-151	F-4J	NF 2xx
VA-93	A-7E	NF 3xx
VA-56	A-7E	NF 4xx
VA-115	A-6E	NF 5xx
	KA-6D	NF 52x
VAW-115	E-2B	NF 60x
VMFP-3 Det 1	RF-4B	RF 61x
VAQ-136 (Feb 80)	EA-6B	NF 62x
HC-1 Det 2	SH-3G	NF 72x

**Commander
Carrier Air Wing Five**
Captain Roger P. Flower

CVW-5
24 February 1981 – 5 June 1981
Indian Ocean/North Arabian Sea
CO - Captain Robert S. Owens
CAG - Captain Roger P. Flower

VF-161	F-4S	NF 1xx
VF-151	F-4S	NF 2xx
VA-93	A-7E	NF 3xx
VA-56	A-7E	NF 4xx
VA-115	A-6E	NF 5xx
	KA-6D	NF 52x
VAW-115	E-2B	NF 60x
VMFP-3 Det A	RF-4B	RF 61x
VAQ-136	EA-6B	NF 62x
HC-1 Det 2	SH-3G	NF 72x

VA-56 *Champs* and VA-93 *Ravens* A-7Es stacked on the starboard bow of *Midway* during Operations in the South China Sea in February 1980. *CDR Pete Clayton*

In the early evening of 29 July 1980, while en route to the Indian Ocean, *Midway* collided with the Panamanian motor vessel *Cactus*. The collision caused strike damage to two VF-151 F-4Js and lesser damage to three other F-4Js, an RF-4B and an EA-6B. The fresnel lens was destroyed and the forward oxygen-nitrogen plant was severely damaged. USN

Above: A VA-115 *Eagles* A-6E TRAM on approach to Atsugi, 29 December 1980. The A-6E was the final production model of the *Intruder*. Externally identical to the A-6A, the A-6E had an ASQ-133 solid-state digital computer. In addition, an APQ-148 multimode-radar replaced the separate search and track radars of the A-6A. Grumman originally proposed that a built-in cannon armament be added, but this suggestion was rejected by the Navy. The A-6E was powered by two 9,300 lb. thrust Pratt & Whitney J52-P-8B turbojets. A-6A BuNo 155673 was modified as the development aircraft for the A-6E, and flew for the first time 27 February 1970. After completion of the 482nd A-6A in December 1970, Grumman changed production to the A-6E version. The first fleet squadron to get the A-6E was VA-85, 9 December 1971. 240 A-6As were rebuilt to the A-6E configuration. The first Target Recognition Attack Multi-sensor (TRAM) A-6E was delivered 14 December 1978. The heart of this system was the AAS-33 TRAM turret, which provided the ability to detect and track targets by their infrared emission, and also allowed the aircraft to self-designate targets for attack by laser-guided weapons. The turret housed a laser and infra-red targeting system with a laser designator, laser rangefinder, and forward-looking infrared (FLIR) sensor. The system enabled the aircraft to identify targets and make an an attack in either day or night.

The TRAM A-6E was also fitted with an APQ-156 radar in place of the APQ-148. The APQ-156 provided search, ground-mapping, tracking and ranging of both fixed and moving targets. Once a target image was locked into the weapon system, the turret's sensors aimed at the locked-in point even as the aircraft maneuvered. Ultimately 228 A-6Es were converted to TRAM configuration. Famous footage from *Desert Storm* depicts a TRAM view of an A-6E attack on a Iraqi bridge. In what General Norman Schwartzkopf described as "the luckiest man in Iraq," a truck just crosses the bridge immediately before a laser-guided bomb hit. In 1988, 62 A-6Es had to be grounded and 119 more were restricted to maneuvers below 3 gs because of wing fatigue. The Navy elected to rewing the A-6E fleet with a composition graphite/epoxy/titanium/aluminum wing manufactured by Boeing-Wichita. The new wing was largely made of composite material, but the control surfaces were still aluminum. The first composite wing aircraft was delivered in October 1990. By January 1995, 85 percent of the A-6E fleet had been re-winged. The A-6E's maximum speed was 644 mph at sea level. It's stall speed was 142 mph. It's empty weight was 26,800 lbs. or 60,400 lbs. at max gross. A maximum ordnance load of 15,000 lbs. could be carried on four underwing and one centerline hardpoints. USN

Above: VRC-50 C-2A *Greyhound* traps aboard *Midway* in February 1980. The C-2 is a derivative of the E-2 *Hawkeye* and replaced the piston-engined C-1 *Trader* in the Carrier Onboard Delivery role. The C-2 shares wings and power plants with the E-2, but has a widened fuselage with a rear loading ramp. The first of two prototypes flew in 1964. The original C-2A aircraft were overhauled to extend their operational life in 1973. In 1984, the Navy bought 39 new C-2As to replace older airframes. All the older C-2As were phased out by 1987, and the last of the new airframes were delivered in 1990. Powered by two Allison T56-425 turboprop engines of 4,800 shp each, the C-2 can deliver a payload of up to 10,000 lbs.

The C-2A's open-ramp flight capability allows airdrop of supplies and personnel from a carrier-launched aircraft. This, plus folding wings and an onboard auxiliary power unit for engine starting and ground power self-sufficiency in remote areas, provide an operational versatility found in no other cargo aircraft. The C-2 crew is two pilots and two aircrewmen. The *Greyhound* can carry 26 passengers or 12 litter patients. The C-2's empty weight is 33,746 lb, normally loaded is 49,394 lb and max takeoff is 60,000 lb. Its maximum speed is 343 knots at 12,000 ft. but normally cruises at 251 knots at 28,700 ft. The maximum rate of climb is 2,610 ft/min. The *Greyhound* stalls at 82 knots and has a range of 1,300 nm. *CDR Pete Clayton*

Midway's MK7 Mod 3 arresting gear easily accommodated this *Enterprise Tomcat*.

Midway's C-13 catapult effortlessly launches VF-114 *Tomcat* in military power.

VF-213 *Black Lion* F-14A *Tomcat* from *Enterprise* is launched on the morning of 30 September 1982, following an emergency recovery aboard *Midway* the previous day. This fighter as well as a VF-114 *Tomcat* and a VA-95 tanker were unable to recover aboard *Enterprise* due to zero-visibility weather conditions in the Northern Pacific during multi-carrier operations. *USN*

Marine Corps RF-4B reconnaissance *Phantom*s were assigned to CVW-5 for 10 years, 1974 to 1984. No other carrier air wing was ever assigned RF-4s. The last RF-4 cruise in 1984 was not without excitement, however. While operating in the North Arabian Sea 25 January 1984, CAPT Michael Healey and his RSO, MAJ John Yencha, Jr., experienced a nose gear hydraulic failure – the nose gear would not come down. In this condition, a barrier arrestment was the best choice to recover the stricken *Phantom*. CAPT Healey flew a flawless approach into the barrier. The RF-4 came to a halt with minimal damage to the aircraft. Neither crewmember was injured. It was *Midway*'s first barrier arrestment since 1976. USN

Speargun 616 an SH-3H *Sea King* assigned to HS-12 flies near Mt. Fujiyama, c. 1985. *Sea Kings* operated from *Midway* from 1970 until she was decommissioned in 1991. HS-12 replaced HC-1 Det 2 in July 1984. The HC-1 Det operated three SH-3Gs in a primary SAR role aboard *Midway*. HS-12 brought six SH-3Hs aboard to provide organic ASW capability as well as to perform SAR and logistics missions. USN

Commander
Carrier Air Wing Five
CDR Larry J. Vernon

CVW-5
14 September 1982 – 10 December 1982
Northern Pacific
CO - Captain Robert S. Owens
CAG - CDR Larry J. Vernon

VF-161	F-4S	NF 1xx
VMFP-3 Det B	RF-4B	RF 11x
VF-151	F-4S	NF 2xx
VA-93	A-7E	NF 3xx
VA-56	A-7E	NF 4xx
VA-115	A-6E	NF 5xx
	KA-6D	NF 52x
VAW-115	E-2B	NF 60x
VAQ-136	EA-6B	NF 60x
HC-1 Det 2	SH-3G	NF 61x

Commander
Carrier Air Wing Five
Captain L. Robert Canepa

CVW-5
25 February 1983 – 9 May 1983
Northern Pacific
CO - Captain Charles R. McGrail
CAG - CDR Larry J. Vernon
CAG - Captain L. Robert Canepa

VF-161	F-4S	NF 1xx
VMFP-3 Det C	RF-4B	RF 1x
VF-151	F-4S	NF 2xx
VA-93	A-7E	NF 3xx
VA-56	A-7E	NF 4xx
VA-115	A-6E	NF 5xx
	KA-6D	NF 52x
VAW-115	E-2B	NF 60x
HC-1 Det 2	SH-3G	NF 61x

CVW-5
28 December 1983 – 23 May 1984
Indian Ocean/North Arabian Sea
CO - Captain Charles R. McGrail
CO - Captain Harry P. Kober, Jr.
CAG - Captain L. Robert Canepa

**Commander
Carrier Air Wing Five**
Captain L. Robert Canepa

VF-161	F-4S	NF 1xx
VMFP-3(to Jul 84)	RF-4B	RF 11x
VF-151	F-4S	NF 2xx
VA-93	A-7E	NF 3xx
VA-56	A-7E	NF 4xx
VA-115	A-6E	NF 5xx
	KA-6D	NF 52x
VAW-115	E-2B	NF 60x
VAQ-136	EA-6B	NF 60x
HC-1 Det 2	SH-3G	NF 61x

The *Grumman* EA-6B *Prowler* is a electronic warfare aircraft designed to jam and deceive enemy radar and communications facilities. The EA-6B is built around the ALQ-99 Tactical Jamming System (TJS) developed by the Airborne Instruments Laboratory Division of Cutler-Hammer. The ALQ-99 system consists of receivers mounted inside a large canoe-shaped fairing on top of the vertical stabilizer that can sense threats in four specific frequency bands. The *Prowler* has a crew of four. The pilot sits in the left front seat. The TJS operator is in the right front seat and doubles as the navigator and as communication operator. The right rear seat position also operated the TJS, while the left rear position is responsible for operating the communications-jamming using the ALQ-92. Development trials were carried out in 1970, including carrier qualifications aboard *Midway*. The first production EA-6B (BuNo 158029) flew for the first time in November 1970. Starting with the 22nd EA-6B, the 9,300 lbs. thrust J52-P-8As were replaced by 11,200 lbs. thrust J52P-408 turbojets. P-408 engines were later retrofitted into all but the first five EA-6Bs. The first EA-6Bs were delivered to VAQ-132 in July 1971. The squadron deployed in *America* to the Gulf of Tonkin and entered combat for the first time in July 1972 along with VAQ-131

aboard *Enterprise*. The two *Prowler* squadrons completed 720 combat sorties over North Vietnam. The ICAP II aircraft has a much more capable system. The last pre-production EA-6B served as the development aircraft for the ICAP II and flew for the first time on 24 June 1980. The TJS was upgraded to the ALQ-99D configuration to cover a wider frequency range. The ASQ-113 jamming system replaced the ASQ-191. In earlier *Prowlers*, the jamming pods each generated signals within one frequency band and were not capable of being reconfigured in flight, but in the ICAP II the pods can generate signals in any one of seven frequency bands, and each pod can jam in two different frequency bands simultaneously. In the ICAP configuration, the communication jamming function was relocated to the navigator's position, up front beside the pilot. The radar and other jamming functions were relocated to the two rear crew positions. The EA-6Bs maximum speed is 658 mph at sea level with an initial rate of climb is 8,600 feet/min. It's stall speed is 133 mph and range is 1628 miles. The *Prowler* weighs 32,160 lbs. empty and 65,000 lbs. at maximum gross. An external load of 15,000 lbs. can be carried on four underwing hardpoints and one centerline hardpoint. The latest EA-6Bs have the ability to fire HARMs. *Robert L. Lawson*

Below: VQ-1 VA-3B rolls out during carrier qualifications aboard *MIDWAY* in October 1983. PR 111 (BuNo 142672) was lost at sea with no survivors 23 January 1985 about 125 NNW of Guam. VQ-1 operated EA-3B Dets aboard deployed PACFLT carriers gathering SIGINT data from 1970 until 1987. The A3D-2Q (EA-3B) was a version of the A3D-2 that was intended specifically for the electronic warfare mission. It differed from the A3D-2 in having the bomb bay replaced with a pressurized compartment. This compartment had accommodations for three electronic countermeasures operators and an evaluator, all sitting in port-facing seats. A set of square windows were cut into the fuselage sides. These extra crew members operated numerous receivers, signal analyzers, recorders and direction finders. A radome was fitted to the tip of the vertical tail. There was a long ventral canoe-shaped fairing housing some of the electronic sensors and equipment. Additional antenna and sensors were attached to the sides of the fuselage. The cockpit was reinforced and was pressurized to 7.5 psi differential over flight pressure versus 3.3 psi for the A3D-2. The YA3D-2Q (Bu No 142670) was first flown 10 December 1958. The prototype and the first 11 A3D-2Qs were fitted with standard wings and were delivered with a tail turret fitted. However, the next twelve were delivered with cambered leading edge wings and the dovetailed rear fuselage with no tail turret. Later, all of the earlier A3D-2Qs were retrofitted with these features. In addition, most of the A3D-2Qs had their original pointed nose radome replaced by the flat panel radome that was fitted to late production A3D-2s. A3D-2Q 142672 was modified in 1958-59 as a VIP transport. After the September 1962 it was designated VA-3B. Only one VA-3B was built. In September 1962, A3D-2Qs were redesignated EA-3Bs. *CDR Pete Clayton*

Above: VMFP-3 RF-4B recovers aboard *Midway* in October 1983 while operating in the South China Sea. Marine Corp RF-4Bs operated from *Midway* from 1974 through 1984. Only two were lost. The Navy initially had no interest in a F4H-1P proposal, as it had the photo *Crusader*. However, the F8U-1P lacked the night reconnaissance capability of the Air Force's RF-110A. In February 1963 the Marine Corps ordered nine of an eventual 46 RF-4Bs. The F4H-1P was redesignated RF-4B in September 1962. It was unarmed. The fighter's radar-equipped nose was replaced with a modified nose that was 4 feet 8 inches longer than that of the F-4B. The APQ-72 radar of the F-4B was replaced by the much smaller APQ-99 forward-looking radar which was optimized for terrain avoidance and terrain-following modes, and could also be used for ground mapping. There were three camera bays in the nose. Station 1 carried a forward oblique or vertical KS-87 camera, Station 2 carried a KA-87 low-altitude camera, and Station 3 carried a KA-55A or KA-91 high-altitude panoramic camera. Film could be developed in flight and film cassettes could be ejected at low altitude.

The rear cockpit was configured for a reconnaissance systems operator, with no flight controls. Two chaff/flare dispensers were installed. For night photography, photoflash cartridges could be ejected upward from each side of the aircraft. The first 34 RF-4Bs retained the engines and the airframe of the F-4B, however, the last twelve were built with the large wheels and the modified wing of the F-4J. The J79-GE-8 engines in these late RF-4Bs were replaced by J79-GE-10 engines. The first RF-4B flew 12 March 1965, and deliveries of 46 airframes took place between May 1965 and December 1970. VMCJ-1 at Iwakuni, Japan took its RF-4Bs to Da Nang in October 1966. In Vietnam three RF-4Bs were lost to ground fire and one was destroyed in an operational accident. In 1975, RF-4Bs were upgraded under *Project SURE* (Sensor Update and Refurbishment Effort). The airframe was strengthened and the wiring was entirely replaced. A datalink, a SLAR, and an infrared reconnaissance set were also fitted. VMFP-3 stood down in August 1990, bringing operations of the RF-4B to an end.
CDR Pete Clayton

Above: The final VF-161 F-4 launch, 25 March 1986. USN

CVW-5
10 June 1985 – 14 October 1985
Indian Ocean/North Arabian Sea
CO - Captain Riley D. Mixson
CAG - CDR Timothy R. Beard

VF-161	F-4S	NF 1xx
VF-151	F-4S	NF 2xx
VA-93	A-7E	NF 3xx
VA-56	A-7E	NF 4xx
VA-115	A-6E	NF 5xx
	KA-6D	NF 52x
VAW-115	E-2C	NF 60x
VAQ-136	EA-6B	NF 60x
HS-12(from Jul 84)	SH-3H	NF 61x

**Commander
Carrier Air Wing Five**
CDR Timothy R. Beard

On 25 March 1986 the last F-4 *Phantom* fighter squadrons in the United States Navy were launched from *Midway*. LT Alan S. Cosgrove and RIO Lt Greg Blankenship of VF-151 flew off the final *Phantom,* NF 210 - BuNo 153879. This aircraft is preserved at Alameda, CA aboard *Hornet* (CVS-12). USN

IV - A MODERN *MIDWAY* AND A NEW *AIR WING FIVE*

VFA-192 *Golden Dragon*'s F/A-18A on approach to NAF Atsugi, c 1988.

Commander
Carrier Air Wing Five
Captain Michael L. Bowman

CVW-5
15 October 1987 – 12 April 1988
Indian Ocean/North Arabian Sea
CO - Captain Richard A. Wilson
CAG - Captain Michael L. Bowman
CAG - Captain David L. Carroll

VFA-195	F/A-18A	NF 1xx
VFA-151	F/A-18A	NF 2xx
VFA-192	F/A-18A	NF 3xx
VA-185	A-6E	NF 4xx
	KA-6D	NF 41x
VA-115	A-6E	NF 5xx
	KA-6D	NF 51x
VAW-115	E-2C	NF 60x
VAQ-136	EA-6B	NF 60x
HS-12	SH-3H	NF 61x

Commander
Carrier Air Wing Five
Captain David L. Carroll

CVW-5
15 August 1989 – 11 December 1989
Indian Ocean/North Arabian Sea
CO - Captain Bernard J. Smith
CAG - Captain David L. Carroll
CAG - Captain James M. Burin

VFA-195	F/A-18A	NF 1xx
VFA-151	F/A-18A	NF 2xx
VFA-192	F/A-18A	NF 3xx
VA-185	A-6E	NF 4xx
	KA-6D	NF 41x
VA-115	A-6E	NF 5xx
	KA-6D	NF 51x
VAW-115	E-2C	NF 60x
VAQ-136	EA-6B	NF 60x
HS-12	SH-3H	NF 61x

A VFA-151 *Vigilantes* F/A-18A and a VA-185 *Nighthawks* KA-6D tanker escort Russian *May* ASW patrol aircraft over the Indian Ocean in February 1988. Development of the F/A-18A originated in the Fighter-Attack, Experimental (VFAX) program to procure a multi-role aircraft to replace the F-4 *Phantom II*, A-4 *Skyhawk*, and A-7 *Corsair II*. The first Full-Scale Development (FSD) F/A-18A (BuNo 160775) was rolled out at St Louis 13 September 1978. The first flight took place at Lambert Field, St Louis 18 November 1978, with test pilot Jack E. Krings at the controls. The first production F/A-18A flew 12 April 1980, and entered operational service with VMFA-314 7 January 1983 and with VFA-113 in March 1983, replacing F-4s and A-7Es, respectively. A total of 371 production F/A-18As were built in blocks 4 through 22 before production shifted to the F/A-18C in September 1987. USN.

VFA-151 *Vigilantes* F/A-18A at Mt Fujiyama, c. 1989. The F/A-18 has outstanding maneuverability due to its excellent thrust to weight ratio (approx .95:1), digital fly-by-wire control system, and leading edge extensions (LEX). The LEX allow the *Hornet* to remain controllable at high angles of attack. The LEX produce strong vortices over the wings, creating turbulent airflow that delays the aerodynamic separation responsible for stall, enabling the *Hornet's* wings to generate lift several times the aircraft's weight, despite high angles of attack. The *Hornet* is thus capable of extremely tight turns over a wide range of speeds. Other design characteristics that enable the *Hornet's* excellent high angle-of-attack performance include very large horizontal stabilators, oversized trailing edge flaps that operate as flaperons, large full-length leading-edge flaps, and a flight control computer that multiplies the movement of each control surface at low speeds and moves the vertical rudders inboard instead of simply left and right. The *Hornet* was among the first aircraft to heavily employ multi-function displays which at the touch of a button allow the pilot to perform either fighter or attack roles or both.

The *Hornet* was the first Navy aircraft to incorporate a digital multiplex avionics bus, enabling easy upgrades. The General Electric F404-GE-400 engines powering the F/A-18A were innovative in that they were designed with reliability and maintainability first. The result is an engine that, while not exceptional on paper in terms of rated performance, demonstrates exceptional robustness and is resistant to stall and flame-out. By contrast, the Pratt & Whitney TF30 engines that powered the F-14A were notoriously prone to flameout in some flight regimes. The engine air inlets of the *Hornet* are fixed, while those of the F-4, F-14, and F-15 have variable geometry or variable ramp engine air inlets. Variable geometry enables high-speed aircraft to keep the velocity of the air reaching the engine below supersonic. Instead, the *Hornet* uses bleed air vents on the inboard surface of the engine air intake ducts to slow and reduce the volume of air reaching the engine. While not as effective as variable geometry, the bleed air technique functions well enough to achieve near Mach 2 speeds. The less sophisticated design is also more robust. USN

VFA-192 *Golden Dragons* F/A-18A configured with HARMs for a SAM suppression mission early in *Operation Desert Storm*. Designed as a light multi-role aircraft, the F/A-18A has a relatively low internal fuel fraction. Its internal fuel capacity is small relative to its take-off weight, at around 23 per cent. Most aircraft in its class have a fuel fraction between .30 to .35. This shortcoming was exacerbated by the addition of new avionics over its lifespan, further reducing the fuel fraction. A total of 1,670 gallons of fuel can be carried internally but three external 330 gallon drop tanks can be fitted, raising total fuel to 2,660 gallons. It's combat radius is 460 miles in a fighter role and 330 miles in a strike configuration. The F/A-18A's maximum range in a ferry configuration is 2,070 miles, without ordnance. *courtesy Captain Jim Burin*

CVW-5
2 October 1990 – 17 April 1991
Arabian Gulf/Operation Desert Storm
CO - Captain Arthur K. Cebrowski
CAG - Captain James M. Burin

VFA-195	F/A-18A	NF 1xx
VFA-151	F/A-18A	NF 2xx
VFA-192	F/A-18A	NF 3xx
VA-185	A-6E	NF 4xx
	KA-6D	NF 41x
VA-115	A-6E	NF 5xx
	KA-6D	NF 51x
VAW-115	E-2C	NF 60x
VAQ-136	EA-6B	NF 60x
HS-12	SH-3H	NF 61x

Commander
Carrier Air Wing Five
Captain James M. Burin

Operation Desert Storm - Golden Dragons VFA-192 F/A-18A on the bow in a pack of *Intruders* and *Hornets*, at the end of a recovery, early in the 1991 Gulf War. *CAPT Scott Dalke, USMC.*

VFA-151 *Vigilantes* and VFA-192 *Golden Dragons Hornets* are being readied for launch during during the early phases of *Operation Desert Storm*. The F/A-18A is powered by twin General Electric F404-GE-400 turbofans, each rated at 10,600 lb. thrust and 15,800 lb. in afterburner. The F/A-18A can reach Mach 1.8 (1190 mph) at 35,000 feet and has a maximum rate of climb is 45,000 ft/m. Its approach speed is 134 knots. The *Hornet* weighs 24,700 lb. empty, 37,150 lb. loaded and max gross takeoff weight is 51,900 lb. A total of 13,700 lbs. of fuel, missiles, and ordnance can be carried. External stores are carried on two wingtip stations for *Sidewinders*, two outboard wing stations for air-to-air or air-to-ground weapons, two inboard wing stations for fuel tanks, air-to-air, or air-to-ground weapons, two nacelle fuselage stations for *Sparrows* or sensor pods and one centerline station for fuel or air-to-ground weapons. The ordnance mix can include two AIM-9 *Sidewinder* air-to-air missiles as well as up to four

AIM-7 *Sparrow* air-to-air missiles. AGM-45 Shrikes, AGM-65 Mavericks, and AGM-88 HARMs can also be carried. The anti-ship mission is expanded by the capability to carry the AGM-84 Harpoon. The *Hornet* is nuclear weapons capable and can deliver mines as well. One 20mm M61A1 cannon with 578 rounds is internally mounted in the nose. F/A-18 pilots were credited with two kills during the Gulf War, both MiG-21s. On the first day of the Gulf War, LT Nick Mongilio and LCDR Mark Fox were launched from *Saratoga* in the Red Sea to bomb an airfield in southwestern Iraq. While en route they were warned by an E-2 of approaching MiG-21 aircraft. The *Hornets* shot down two MiGs and resumed their bombing run, each delivering four 2,000 lb bombs, before returning to *Saratoga*. Mongilio and Fox become the first pilots to register air-to-air kills while still completing their original air-to-ground mission. *CAPT Scott Dalke, USMC.*

Nighthawks KA-6D tanker, one of two assigned to VA-185, is launched from the port catapult during *Operation Desert Storm*. In September 1987, VA-185, a second A-6E squadron, was assigned to CVW-5. The *Nighthawks* moved into Ready Room 4. VA-185 and VA-115 were each assigned seven A-6E aircraft and two KA-6D tankers bringing the total number of *Intruders* aboard to 18. The two A-6 squadron air wing composition was a project of Secretary of the Navy John Lehman, a reserve bombardier/navigator and a leading proponent of the *Intruder*. During his tenure as SecNav under President Reagan, five carrier air wings were expanded to include two squadrons of the Grumman all-weather attack aircraft – CVW-2 in *Ranger*, CVW-3 in *John F. Kennedy*, CVW-5 in *Midway*, CVW-8 in *Theodore Roosevelt* and CVW-13 in *Coral Sea*. USN

USS Midway Aircraft Complements

1945 - 1948

VF	VB	VF(P)					Total
64	64	4					132

1949 - 1950

VF	VA	VC(AW)	VC(AEW)	VC(P)	HU	Total
48	32	4	4	4	2	94

1950 - 1952

VF	VA	VC(AW)	VC(AEW)	VC(P)	HU	Total
64	16	4	4	4	2	94

1953 - 1955

VF	VA	VC(H)	VC(AW)	VC(AEW)	VC(P)	HU	Total
60	14	3	4	4	4	2	91

1958 - 1965

VF	VA	VA(M)	VAH	VAW	VA(Q)	VFP	HU	Total
20	20	12	10	3	3	3	2	73

1971 - 1972

VF	VA	VA(M)	VAW	VFP	VAQ	HC	Total
24	24	14	4	3	3	3	75

1973 - 1974

VF	VA	VA(M)	VAW	VFP	HC	Total
24	24	14	4	3	3	72

1975 - 1977

VF	VA	VA(M)	VAW	VMAQ	VMFP	HC	Total
22	24	14	4	3	3	3	73

1978 - 1984

VF	VA	VA(M)	VAW	VAQ	VMFP	HC	Total
20	24	14	4	4	3	3	72

1984 - 1985

VF	VA	VA(M)	VAW	VAQ	HS	Total
20	24	14	4	4	6	72

1987 - 1991

VFA	VA(M)	VAW	VAQ	HS	Total
36	18	4	4	6	68

LAST LANDING AND FINAL LAUNCH

Midway got underway under her own power for the last time 24 September 1991. She recovered and launched her last aircraft on that day. The last trap was made by Captain Patrick Moneymaker, *Commander Carrier Air Wing 14*, in a VFA-151 *Vigilantes* F/A-18A *Hornet*.

USS Midway General Aviation Dimensions

	As Built 1945	SCB-110 1957	SCB-101.66 1970	Blister 1986
Length of the flight deck	924 feet	977 feet	972 feet	972 feet
Length of the angle deck		531 feet	651 feet	651 feet
Hangar deck	692 x 95 feet	628 feet	628 feet	560 feet

Aviation Launch and Recovery Equipment

Catapults

1945 - 1955	two H4-1 - capacity: 28,000 lbs @ 90 mph
1958 - 1965	two C11-1- capacity: 39,000 lbs @ 156 mph and one C11-2 all with dry receivers -
1970 - 1991	two C13 with wet receivers - capacity: 70,000 lbs @ 160 knots

Arresting Gear

1945 - 1955	Mk 5-0
1958 - 1965	seven Mk 7 mod 1 (five cross deck pendants plus two barricades)
1970 - 1991	four Mk 7 mod 3 (three cross deck pendants plus one barricade)

Aviation Fuel

1945 - 1955	332,000 gallons Av Gas
1958 - 1965	366,000 gallons Av Gas 873,000 gallons JP5
1970 - 1991	1,220,000 gallons JP-5

This VF-21 F-4B (BuNo 152219) crewed by LT Jack Batson and LT Robert Doremus shot down a MiG-17 with an AIM-7 *Sparrow* missile 17 June 1965. A second kill was later credited to this crew for a MiG-17 which crashed after ingesting parts from the explosion of the first MiG.

This VF-161 F-4B (BuNo 153915) crewed by LT P Arwood and LT Jim Bell shot down a MIG-17 18 M 1972. BuNo 153915 is preserved at the Naval Aviati Museum, Pensacola, FL. *CDR Rob Anderson, USNR (R*

USS Midway MiG Kill Summary

Date	Aircraft	BuNo	Squadron	Modex	Crew	Aircraft	Weapon
17 Jun 65	F-4B	151488	VF-21	NE 101	Cdr Lou Page LT John Smith	MiG-17	AIM-7
17 Jun 65	F-4B	152219	VF-21	NE 102	LT Jack Batson LT Robert Doremus	MiG-17	AIM-7
20 Jun 65	A-1H	139768	VA-25	NE 577	LT Clint Johnson	MiG-17	20mm
	A-1H	142070	VA-25	NE 573	LTjg Charles Hartman		20mm
18 May 72	F-4B	153068	VF-161	NF 110	LT Hank Bartholomey LT Oran Brown	MiG-19	AIM-9
18 May 72	F-4B	153915	VF-161	NF 105	LT Pat Arwood LT Jim Bell	MiG-19	AIM-9
23 May 72	F-4B	153020	VF-161	NF 100	LCDR Mugs McKeown LT Jack Ensch	MiG-17	AIM-9
23 May 72	F-4B	153020	VF-161	NF 100	LCDR Mugs McKeown LT Jack Ensch	MiG-17	AIM-9
12 Jan 73	F-4B	153045	VF-161	NF 102	LT Vic Koveleski LTjg Jim Wise	MiG-17	AIM-9